CAN BUS INTERVIEW QUESTIONS:

1. What is CAN and its uses?
Answer: 1. CAN is a multi-master broadcast serial bus standard for connecting electronic control unit (ECUs).

2. Controller–area network (CAN or CAN-bus) is a vehicle bus standard designed to allow microcontrollers and devices to communicate with each other within a vehicle without a host computer.

3. CAN is a message-based protocol, designed specifically for automotive applications but now also used in other areas such as industrial automation and medical equipment.

4. The Controller Area Network (CAN) bus is a serial asynchronous bus used in instrumentation applications for industries such as automobiles.

USES:
– More reliably, e.g., fewer plug-in connectors that might cause errors.
– Wiring less complicated, more economic.
– Easy to implement, changes, too.
– Additional elements (e.g., control units) are easy to integrate.
– Installation place exchangeable without electric problems.
– Wire may be diagnosed.

2. CAN frame works?
Answer: SOF – 1 Dominant
Arbitration Field – 11 bit Identifier, 1 bit RTR (or) 11 bit, 1SRR, 1IDE, 18 bit, 1RTR

Control Field – IDE, r0, 4 bits (DLC)

Data Field – (0-8) Bytes

CRC Field – 15 bits, Delimiter (1 bit recessive)

ACK Field – 1 bit, Delimiter (1 bit recessive)

EOF – 7 bits recessive

IFS – 3 bits recessive

Types of frames – Data, remote, Error frame and Overload frame

Types of errors – ACK error, Bit error, Stuff error, Form error, CRC error

Error frame – 0-12 superposition flags, 8 recessive (Delimiter)

Overload frame – 0-12 superposition flags, 8 recessive (Delimiter)

3. Why CAN is having 120 ohms at each end?
Answer: To minimize the reflection reference, to reduce noise. To ensure that reflection does not cause communication failure, the transmission line must be terminated.

4. Why CAN is message oriented protocol?
Answer: CAN protocol is a message-based protocol, not an address based protocol. This means that messages are not transmitted from one node to another node based on addresses. Embedded in the CAN message itself is the priority and the contents of the data being transmitted. All nodes in the system receive every message transmitted on the bus (and will acknowledge if the message was properly received). It is up to each node in the system to decide whether the message received should be immediately discarded or kept to be processed. A single message can be destined for one particular node to receive, or many nodes based on the way the network and system are designed. For example, an automotive airbag sensor can be connected via CAN to a safety system router node only. This router node takes in other safety system information and routes it to all other nodes on the safety system network. Then all the other nodes on the safety system network can receive the latest airbag sensor information from the router at the same time, acknowledge if the message was received properly, and decide whether to utilize this information or discard it.

5. CAN logic what it follows?
Answer: Wired AND logic

6. What is CAN Arbitration?
Answer: CAN Arbitration is nothing but the node trying to take control on the CAN bus.

7. How CAN will follow the Arbitration?
Answer: CSMA/CD + AMP (Arbitration on Message Priority)
Two bus nodes have got a transmission request. The bus access method is CSMA/CD+AMP (Carrier Sense Multiple Access with Collision Detection and Arbitration on Message Priority). According to this algorithm both network nodes wait until the bus is free (Carrier Sense). In that case the bus is free both nodes transmit their dominant start bit (Multiple Access). Every bus node reads back bit by bit from the bus during the complete message and compares the transmitted value with the received value. As long as the bits are identical from both transmitters nothing happens. The first time there was a difference – in this example the 7th bit of the message – the arbitration process takes place: Node A transmits a dominant level, node B transmits a recessive level. The recessive level will be overwritten by the dominant level. This is detected by node B because the transmitted value is not equal to the received value (Collision Detection). At this point of time node B has lost the arbitration, stops the transmission of any further bit immediately and switches to receive mode, because the message that has won the arbitration must possibly be processed by this node (Arbitration on Message Priority)
For example, consider three CAN devices each trying to transmit messages:

• Device 1 – address 433 (decimal or 00110110001 binary)
• Device 2 – address 154 (00010011010)
• Device 3 – address 187 (00010111011)

Assuming all three see the bus is idle and begin transmitting at the same time, this is how the arbitration works out. All three devices will drive the bus to a dominant state for the start-of-frame (SOF) and the two most significant bits of each message identifier. Each device will monitor the bus and determine success. When they write bit 8 of the message ID, the device writing message ID 433 will notice that the bus is in the dominant state when it was trying to let it be recessive, so it will assume a collision and give up for now. The remaining devices will continue writing bits until bit 5, then the device writing message ID 187 will notice a collision and abort transmission. This leaves the device writing message ID 154 remaining. It will continue writing bits on the bus until complete or an error is detected. Notice that this method of arbitration will always cause the lowest numerical value message ID to have priority. This same method of bit-wise arbitration and prioritization applies to the 18-bit extension in the extended format as well.

8.   What is the speed of CAN?
Answer: 40m @1Mbps and if the cable length increases will decrease the speed, due to RLC on the cable.

9.   If master sends 764 and Slave sends 744 which will get the arbitration?
Answer: Starts from MSB, first nibble is same, Master sends 7, slaves also sends 7 the message with more dominant bits will gain the arbitration, lowest the message identifier higher the priority.

10.   Standard CAN and Extended CAN difference?
Answer: Number of identifiers can be accommodated for standard frame are 2power11.
Number of identifiers more compare to base frame, for extended frame are 2power29.
IDE bit – 1 for extended frame.

IDE bit – 0 for Standard frame.

11.   What is bit stuffing?
Answer: CAN uses a Non-Return-to-Zero protocol, NRZ-5, with bit stuffing. The idea behind bit stuffing is to provide a guaranteed edge on the signal so the receiver can resynchronize with the transmitter before minor clock discrepancies between the two nodes can cause a problem. With NRZ-5 the transmitter transmits at most five consecutive bits with the same value. After five bits with the same value (zero or one), the transmitter inserts a stuff bit with the opposite state.

12. What is the use of bit stuffing?
Answer: Long NRZ messages cause problems in receivers
• Clock drift means that if there are no edges, receivers lose track of bits

• Periodic edges allow receiver to resynchronize to sender clock

13.   What are the functions of CAN transceiver?
Answer: The transceiver provides differential transmit capability to the bus and differential receive capability to the CAN controller. Transceiver provides an advanced interface between the protocol controller and the physical bus in a Controller Area Network (CAN) node. Typically, each node in a CAN system must have a device to convert the digital signals generated by a CAN controller to signals suitable for transmission over the bus cabling

(differential output). It also provides a buffer between the CAN controller and the high-voltage spikes that can be generated on the CAN bus by outside sources (EMI, ESD, electrical transients, etc.).

The can transceiver is a device which detects the signal levels that are used on the CAN bus to the logical signal levels recognized by a microcontroller.

14.   Functionality of Data link layer in CAN?
Answer: LLC (Logical Link Control) – Overload control, notification, Message filtering and Recovery management functions.
MAC (Medium Access Control) – Encapsulation/ de-capsulation, error detection and control, stuffing and de-stuffing and serialization/de-serialization.
15.   What is meant by synchronization?
Answer: Synchronization is timekeeping which requires the coordination of events to operate a system in unison.
16.   What is meant by Hard synchronization and soft synchronization?
Answer: Hard Synchronization to be performed at every edge from recessive-to-dominant edge during Bus Idle. Additionally, Hard Synchronization is required for each received SOF bit. An SOF bit can be received both during Bus Idle, and also during Suspend Transmission and at the end of Interframe Space. Any node disables Hard Synchronization if it samples an edge from recessive to dominant or if it starts to send the dominant SOF bit.
Two types of synchronization are supported:

– **Hard synchronization** is done with a falling edge on the bus while the bus is idle, which is interpreted as a Start of frame (SOF). It restarts the internal Bit Time Logic.
– **Soft synchronization** is used to lengthen or shorten a bit time while a CAN frame is received.
17.   What is the difference between function and physical addressing?
Answer: Functional addressing is an addressing scheme that labels messages based upon their operation code or content. Physical addressing is an addressing scheme that labels messages based upon the physical address location of their source and/or destination(s).
18.   What happens if I have to send more than 8-bytes of data?
Answer: The J1939 standard has defined a method of communicating more than 8 bytes of data by sending the data in packets as specified in the Transport Protocol (TP). There are two types of TP, one for broadcasting the data, and the other for sending it to a specific address. DTC consists of 4 components – SPN, FMI, OC and CM.

A DTC is a combination of four independent fields: the Suspect Parameter Number (SPN) of the channel or feature that can have faults; a Failure Mode Identifier (FMI) of the specific fault; the occurrence count (OC) of the SPN/FMI combination; and the SPN conversion method (CM) which tells the receiving mode how to interpret the SPN. Together, the SPN, FMI, OC and CM form a number that a diagnostic tool can use to understand the failure that is being reported.

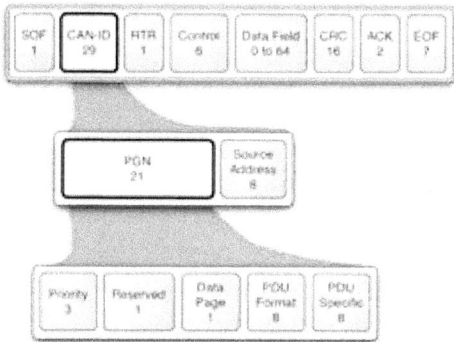

Figure 1 – Message format

19. What is KWP2000?

Answer: KWP 2000(ISO14230) is a Diagnostic communications standard. Specifies possible system configurations using the K & L lines. As 9141-2 but limited to the physical characteristics. Specifies possible system configurations using the K & L lines.

- 5 Baud wake up as 9141- 2
- New fast initialisation method

20. What is OBDII?

Answer: On-Board Diagnostics in an automotive context is a generic term referring to a vehicle's self-diagnostic and reporting capability

21. Why Diagnostic Standards?

Answer: As systems got more complex the link between cause and symptom became less obvious. This meant that electronic systems had to have some level of self diagnosis and to communicate to the outside world. Initially many systems used their own protocols which meant that garages had to have a large number of tools – even to diagnose a single vehicle.

22. What is meant by verification and validation??

Answer: Verification and Validation (V&V) is the process of checking that a software system meets specifications and that it fulfills its intended purpose. It is normally part of the software testing process of a project.

According to the Capability Maturity Model (CMMI-SW v1.1),

- Verification: The process of evaluating software to determine whether the products of a given development phase satisfy the conditions imposed at the start of that phase.
- Validation: The process of evaluating software during or at the end of the development process to determine whether it satisfies specified requirements.
- **Verification** shows conformance with specification; **validation** shows that the program meets the customer's needs

23. Can you have two transmitters using the same exact header field?

Answer: No – that would produce a bus conflict

• Unless you have middleware that ensures only one node can transmit at a time

– For example use a low priority message as a token to emulate token-passing

24. CAN physical layer voltage levels

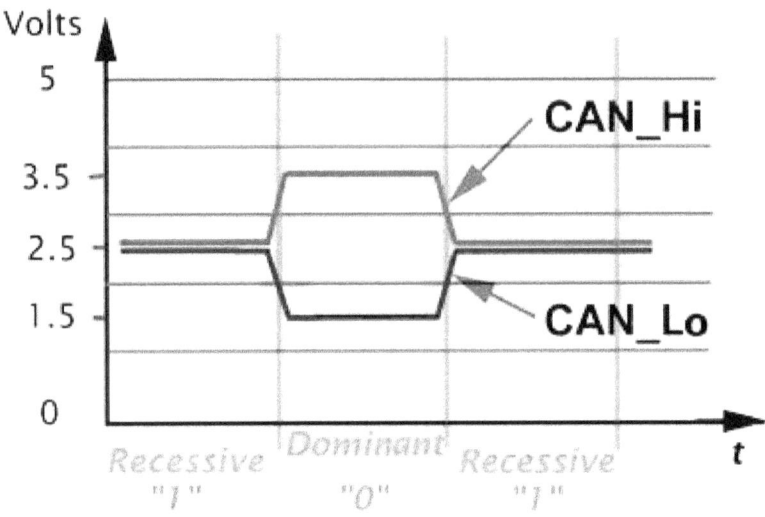

25. CAN bit timing:

According to the CAN specification, the bit time is divided into four segments. The Synchronization Segment, the Propagation Time Segment, the Phase Buffer Segment 1, and the Phase Buffer Segment 2. Each segment consists of a specific, programmable number of time quanta (see Table 1). The length of the time quantum (tq), which is the basic time unit of the bit time, is defined by the CAN controller's system clock fsys and the Baud Rate Prescaler (BRP) :tq = BRP / fsys. Typical system clocks are : fsys = fosc or fsys = fosc/2.

The Synchronization Segment Sync_Seg is that part of the bit time where edges of the CAN bus level are expected to occur; the distance between an edge that occurs outside of Sync_Seg and the Sync_Seg is called the phase error of that edge. The Propagation Time Segment Prop_Seg is intended to compensate for the physical delay times within the CAN network. The Phase Buffer Segments Phase_Seg1 and Phase_Seg2 surround the Sample Point. The (Re-)Synchronization Jump Width (SJW) defines how far a resynchronization may move the Sample Point inside the limits defined by the Phase Buffer Segments to compensate for edge phase errors.

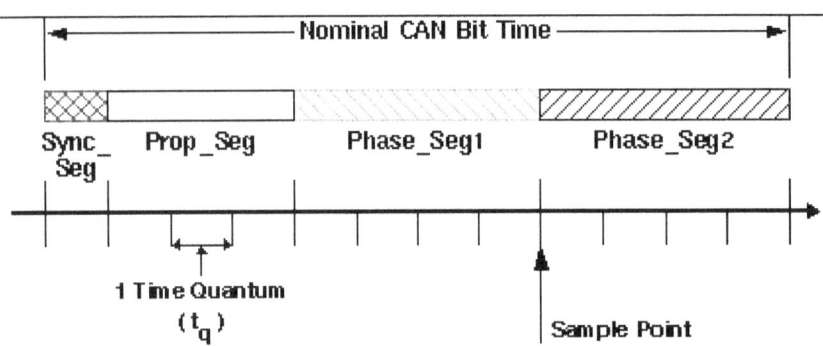

Two types of synchronization exist : Hard Synchronization and Resynchronization. A Hard Synchronization is done once at the start of a frame; inside a frame only Resynchronizations occur.

• **Hard Synchronization** After a hard synchronization, the bit time is restarted with the end of Sync_Seg, regardless of the edge phase error. Thus hard synchronization forces the edge which has caused the hard synchronization to lie within the synchronization segment of the restarted bit time.

• **Bit Resynchronization** Resynchronization leads to a shortening or lengthening of the bit time such that the position of the sample point is shifted with regard to the edge.

26. Formula for Baudrate calculation?

The baud rate is calculated as:

$$\text{baud rate (bits per second)} = 18.432 \times 10^6 / BRP / (1 + TSEG1 + TSEG2)$$

27. What happen when two CAN nodes are sending same identifier at a same time?

Two nodes on the network are not allowed to send messages with the same id. If two nodes try to send a message with the same id at the same time arbitration will not work. Instead, one of the transmitting nodes will detect that his message is distorted outside of the arbitration field. The nodes will then use the error handling of CAN, which in this case ultimately will lead to one of the transmitting node being switched off (bus-off mode).

28. what is the difference between Bit Rate and Baud Rate?

The difference between Bit and Baud rate is complicated and intertwining. Both are dependent and inter-related. But the simplest explanation is that a Bit Rate is how many data bits are transmitted per second. A baud Rate is the number of times per second a signal in a communications channelchanges.Bit rates measure the number of data bits (that is 0's and 1's) transmitted in one second in a communication channel. A figure of 2400 bits per second means 2400 zeros or ones can be transmitted in one second, hence the abbreviation "bps." Individual characters (for example letters or numbers) that are also referred to as bytes are composed of several bits.A baud rate is the number of times a signal in a communications channel changes state or varies. For example, a 2400 baud rate means that the channel can change states up to 2400 times per second. The term "change state" means that it can change from 0 to 1 or from 1 to 0 up to X (in this case, 2400) times per second. It also refers to the actual state of the connection, such as voltage, frequency, or phase level).The main difference between the two is that one change of state can transmit one bit, or slightly more or less than one bit, that depends on the modulation technique used. So the bit rate (bps) and baud rate (baud per second) have this connection:bps = baud per second x the number of bit per baudThe modulation technique determines the number of bit per baud. Here are two examples:When FSK (Frequency Shift Keying, a transmission technique) is used, each baud transmits one bit. Only one change in state is required to send a bit. Thus, the modem's bps rate is equal to the baud rate. When a baud rate of 2400 is used, a modulation technique called phase modulation that transmits four bits per baud is used. So:2400 baud x 4 bits per baud = 9600 bpsSuch modems are capable of 9600 bps operation.

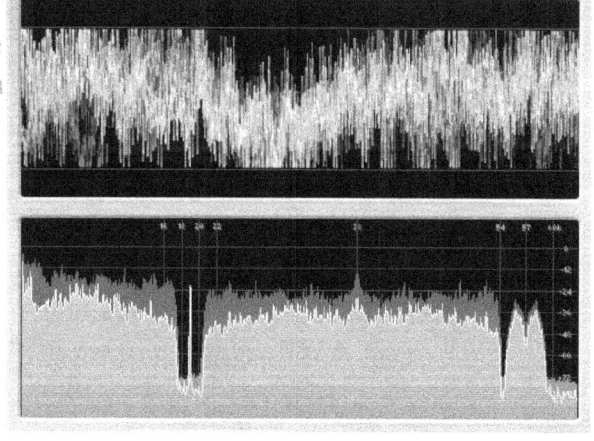

Embedded C

1. **What is the difference between declaration and definition?**
Answer: Definition means where a variable or function is defined in reality and actual memory is allocated for variable or function.
Declaration means just giving a reference of a variable and function.

2. **What are the different storage classes in C?**
Answer: AUTO, STATIC, EXTERN, REGISTER
auto is the default storage class for local variables.

```
{

int Count;

auto int Month;

}
```

register is used to define local variables that should be stored in a register instead of RAM This means that the variable has a maximum size equal to the register size (usually one word) and cannot have the unary '&' operator applied to it (as it does not have a memory location).

```
        {

        register int  Miles;

        }
```

Register should only be used for variables that require quick access – such as counters. It should also be noted that defining 'register' goes not mean that the variable will be stored in a register. It means that it MIGHT be stored in a register – depending on hardware and implementation restrictions.

static – Storage Class
static is the default storage class for global variables. The two variables below
(**count** and **road**) both have a static storage class.

```
static int Count;int
Road;main(){printf("%d\n",
Count);printf("%d\n", Road);}
```

'static' can also be defined within a function. If this is done. the variable is initalized at compilation time and retains its value between calls. Because it is initialized at compilation time, the initialization value must be a constant. This is serious stuff – tread with care.

```
void Func(void){static Count=1;}
```

Here is an example

There is one very important use for 'static'. Consider this bit of code.

```
char *Func(void);main(){char
*Text1;Text1 = Func();}char
*Func(void){char
Text2[10]="martin";return(Text2);}
```

'Func' returns a pointer to the memory location where 'Text2' starts BUT Text2 has a storage class of auto and will disappear when we exit the function and could be overwritten by something else. The answer is to specify:

```
static char Text[10]="martin";
```

The storage assigned to 'Text2' will remain reserved for the duration if the program.

extern – storage Class

extern defines a global variable that is visable to ALL object modules. When you use 'extern' the variable cannot be initalized as all it does is point the variable name at a storage location that has been previously defined.

Source 1   Source 2

_____   _____

extern int count;   int count=5;

write()   main()

{   {

printf("count is %d\n", count);   write();

}   }

Count in 'source 1' will have a value of 5. If source 1 changes the value of count – source 2 will see the new value. Here are some example source files.

3. **What is interrupt?**
 Answer: Interrupts (also known as traps or exceptions in some processors) are a technique of diverting the processor from the execution of the current program so that it may deal with some event that has occurred. Such an event may be an error from a peripheral, or simply that an I/O device has finished the last task it was given and is now ready for another. An interrupt is generated in your computer every time you type a key or move the mouse. You can think of it as a hardware-generated function call.

4. **What is Hardware Interrupt?**
Answer: There are two ways of telling when an I/O device (such as a serial controller or a disk controller) is ready for the next sequence of data to be transferred. The first is busy waiting or polling, where the processor continuously checks the device's status register until

the device is ready. This wastes the processor's time but is the simplest to implement. For some time-critical applications, polling can reduce the time it takes for the processor to respond to a change of state in a peripheral.

5. What is Software Interrupt?

Answer: A software interrupt is generated by an instruction. It is the lowest-priority interrupt and is generally used by programs to request a service to be performed by the system software (operating system or firmware).

Difference between Hardware Interrupt and Software Interrupt

An interrupt is a special signal that causes the computer's central processing unit to suspend what it is doing and transfers its control to a special program called an **interrupt handler**. The responsibility of an interrupt handler is to determine what caused the interrupt, service the interrupt and then return the control to the point from where the interrupt was caused. The difference between **hardware interrupt** and **software interrupt** is as below:

Hardware Interrupt: This interrupt is caused by some external device such as request to start an I/O or occurrence of a hardware failure.

Software Interrupt: This interrupt can be invoked with the help of INT instruction. A programmer triggered this event that immediately stops execution of the program and passes execution over to the INT handler. The INT handler is usually a part of the operating system and determines the action to be taken e.g. output to the screen, execute file etc.

Thus a software interrupt as it's name suggests is driven by a software instruction and a hardware interrupt is the result of external causes.

6. What is Interrupt latency? How do you measure interrupt latency? How to reduce the interrupt latency?

Answer: Interrupt latency is the time between interrupt request and execution of first instruction of the ISR.

We need a oscilloscope or a logic state analyzer. By entering the interrupt service routine (ISR), you need to activate an available port on your hardware (like a led port or so on) and deactivate it just before returning from the ISR. You can do that by writing the appropriate code.

By connecting one input of the oscilloscope (or logic state analyzer) to the INTR pin of the microprocessor and the second one to the port you activate/deactivate, you can measure the latency time and the duration of the ISR

Causes of interrupt latencies

- The first delay is typically in the hardware: The interrupt request signal needs to be synchronized to the CPU clock. Depending on the synchronization logic, typically up to 3 CPU cycles can be lost before the interrupt request has reached the CPU core.
- The CPU will typically complete the current instruction. This instruction can take a lot of cycles; on most systems, divide, push-multiple or memory-copy instructions are the instructions which require most clock cycles. On top of the cycles required by the CPU, there are in most cases additional cycles required for memory access. In an ARM7 system, the instruction STMDB SP!,{R0-R11,LR}; Push parameters and perm. Registers is typically the worst case instruction. It stores 13 32-bit registers on

the stack. The CPU requires 15 clock cycles. The memory system may require additional cycles for wait states.

- After completion of the current instruction, the CPU performs a mode switch or pushes registers (typically PC and flag registers) on the stack. In general, modern CPUs (such as ARM) perform a mode switch, which requires less CPU cycles than saving registers.
- Pipeline fill: Most modern CPUs are pipelined. Execution of an instruction happens in various stages of the pipeline. An instruction is executed when it has reached its final stage of the pipeline. Since the mode switch has flushed the pipeline, a few extra cycles are required to refill the pipeline.

7. Vonneuman and harvard architecture differences?

Answer: The name Harvard Architecture comes from the Harvard Mark I relay-based computer. The most obvious characteristic of the Harvard Architecture is that it has physically separate signals and storage for code and data memory. It is possible to access program memory and data memory simultaneously. Typically, code (or program) memory is read-only and data memory is read-write. Therefore, it is impossible for program contents to be modified by the program itself.

The vonneumann Architecture is named after the mathematician and early computer scientist John von Neumann. von Neumann machines have shared signals and memory for code and data. Thus, the program can be easily modified by itself since it is stored in read-write memory.

Harvard architecture has separate data and instruction busses, allowing transfers to be performed simultaneously on both buses. **Von Neumann architecture** has only one bus which is used for both data transfers and instruction fetches, and therefore data transfers and instruction fetches must be scheduled – they cannot be performed at the same time.

It is possible to have two separate memory systems for a **Harvard architecture**. As long as data and instructions can be fed in at the same time, then it doesn't matter whether it comes from a cache or memory. But there are problems with this. Compilers generally embed data (literal pools) within the code, and it is often also necessary to be able to write to the instruction memory space, for example in the case of self modifying code, or, if an ARM debugger is used, to set software breakpoints in memory. If there are two completely separate, isolated memory systems, this is not possible. There must be some kind of bridge between the memory systems to allow this.

Using a simple, unified memory system together with a Harvard architecture is highly inefficient. Unless it is possible to feed data into both buses at the same time, it might be better to use a von Neumann architecture processor.

Use of caches

At higher clock speeds, caches are useful as the memory speed is proportionally slower. **Harvard architectures** tend to be targeted at higher performance systems, and so caches are nearly always used in such systems.

Von Neumann architectures usually have a single unified cache, which stores both instructions and data. The proportion of each in the cache is variable, which may be a good thing. It would in principle be possible to have separate instruction and data caches, storing data and instructions separately. This probably would not be very useful as it would only be possible to ever access one cache at a time.

Caches for Harvard architectures are very useful. Such a system would have separate caches for each bus. Trying to use a shared cache on a Harvard architecture would be very inefficient since then only one bus can be fed at a time. Having two caches means it is possible to feed both buses simultaneously....exactly what is necessary for a Harvard architecture.

This also allows to have a very simple unified memory system, using the same address space for both instructions and data. This gets around the problem of literal pools and self modifying code. What it does mean, however, is that when starting with empty caches, it is necessary to fetch instructions and data from the single memory system, at the same time. Obviously, two memory accesses are needed therefore before the core has all the data needed. This performance will be no better than a von Neumann architecture. However, as the caches fill up, it is much more likely that the instruction or data value has already been cached, and so only one of the two has to be fetched from memory. The other can be supplied directly from the cache with no additional delay. The best performance is achieved when both instructions and data are supplied by the caches, with no need to access external memory at all.

This is the most sensible compromise and the architecture used by ARMs Harvard processor cores. Two separate memory systems can perform better, but would be difficult to implement

8.   RISC and CISC differences?
Answer:
CISC: (Complex Instruction Set Computer)
Eg: Intel and AMD CPU's
- CISC chips have a large amount of different and complex instructions.
- CISC chips are relatively slow (compared to RISC chips) per instruction, but use little (less than RISC) instructions.
- CISC architecture is to complete a task in as few lines of assembly as possible. This is achieved by building processor hardware that is capable of understanding and executing a series of operations.
- In CISC, compiler has to do very little work to translate a high-level language statement into assembly. Because the length of the code is relatively short, very little RAM is required to store instructions. The emphasis is put on building complex instructions directly into the hardware.
- When executed, this instruction loads the two values into separate registers, multiplies the operands in the execution unit, and then stores the product in the appropriate register.

RISC: (Reduced Instruction Set Computer)
Eg: Apple, ARM processors
- Fewer, simpler and faster instructions would be better, than the large, complex and slower CISC instructions. However, more instructions are needed to accomplish a task.
- RISC chips require fewer transistors, which makes them easier to design and cheaper to produce.
- it's easier to write powerful optimized compilers, since fewer instructions exist.
- RISC is cheaper and faster.

- RISC puts a greater burden on the software. Software needs to become more complex. Software developers need to write more lines for the same tasks. Therefore they argue that RISC is not the architecture of the future, since conventional CISC chips are becoming faster and cheaper
- Simple instructions that can be executed within one clock cycle.
- "MULT" command described above could be divided into three separate commands: "LOAD," which moves data from the memory bank to a register, "PROD," which finds the product of two operands located within the registers, and "STORE," which moves data from a register to the memory banks.
- At first, this may seem like a much less efficient way of completing the operation. Because there are more lines of code, more RAM is needed to store the assembly level instructions. The compiler must also perform more work to convert a high-level language statement into code of this form.
- Separating the "LOAD" and "STORE" instructions actually reduces the amount of work that the computer must perform.
- Major problem of RISC – they don't afford the widespread compatibility, that x86 chips do.

CISC	RISC
Emphasis on hardware	Emphasis on software
Includes multi-clock complex instructions	Single-clock, reduced instruction only
Memory-to-memory: "LOAD" and "STORE" incorporated in instructions	Register to register: "LOAD" and "STORE" are independent instructions
Small code sizes, high cycles per second	Low cycles per second, large code sizes
Transistors used for storing complex instructions	Spends more transistors on memory registers

CISC	RISC
Complex instructions require multiple cycles	Reduced instructions take 1 cycle
Many instructions can reference memory	Only Load and Store instructions can reference memory
Instructions are executed one at a time	Uses pipelining to execute instructions
Few general registers	Many general registers

9.   What are the startup code steps?
Answer:
1. Disable all the interrupts.
2. Copy and initialized data from ROM to RAM.
3. Zero the uninitialized data area.
4. Allocate space and for initialize the stack.
5. Initialize the processor stack pointer
6. Call main

10.   What are the booting steps for a CPU?
Answer:

- The power supply does a self check and sends a power-good signal to the CPU. The CPU starts executing the code stored in ROM on the motherboard starts the address 0xFFFF0.

  - The routines in ROM test the central hardware, search for video ROM, perform a checksum on the video ROM and executes the routines in video ROM.
  - The routines in the mother board ROM then continue searching for any ROM, checksum and executes these routines.
  - After performing the POST (Power On-Self Test) is executed. The system will search for a boot device.
  - Assuming that the valid boot device is found, IO.SYS is loaded into memory and executed.IO.SYS consists primarily of initialization code and extension to the memory board ROM BIOS.
  - MSDOS.SYS is loaded into memory and executed. MSDOS.SYS contains the DOS routines.
  - CONFIG.SYS (created and modified by the user. load additional device drivers for peripheral devices), COMMAND.COM (It is command interpreter- It translates the commands entered by the user. It also contains internal DOS commands. It executes and AUTOEXEC.BAT),AUTOEXEC.BAT (It contains all the commands that the user wants which are executed automatically every time the computed is started).

11. What Little-Endian and Big-Endian? How can I determine whether a machine's byte order is big-endian or little endian? How can we convert from one to another?

First of all, Do you know what Little-Endian and Big-Endian mean?

Little Endian means that the lower order byte of the number is stored in memory at the lowest address. and the higher order byte is stored at the highest address. That is, the little end comes first.

For example, a 4 byte, 32-bit integerByte3 Byte2 Byte1 Byte0will be arranged in memory as follows:Base_Address+0 Byte0

Base_Address+1 Byte1

Base_Address+2 Byte2

Base_Address+3 Byte3Intel processors use "Little Endian" byte order."Big Endian" means that the higher order byte of the number is stored in memory at the lowest address, and the lower order byte at the highest address. The big end comes first.Base_Address+0 Byte3

Base_Address+1 Byte2

Base_Address+2 Byte1

Base_Address+3 Byte0Motorola, Solaris processors use "Big Endian" byte order.In "Little Endian" form, code which picks up a 1, 2, 4, or longer byte number proceed in the same way for all formats. They first pick up the lowest order byte at offset 0 and proceed from there. Also, because of the 1:1 relationship between address offset and byte number (offset 0 is byte 0), multiple precision mathematic routines are easy to code. In "Big Endian" form, since the high-order byte comes first, the code can test whether the number is positive or negative by looking at the byte at offset zero. Its not required to know how long the number is, nor does the code have to skip over any bytes to find the byte containing the sign information. The numbers are also stored in the order in which they are printed out, so binary to decimal routines are particularly efficient. Here is some code to determine what is the type of your machine.

```
int num = 1;
if(*(char *)&num == 1)
{
printf("\nLittle-Endian\n");
}
else
{
printf("Big-Endian\n");
}And here is some code to convert from one Endian to another.int myreversefunc(int num)
{
int byte0, byte1, byte2, byte3;byte0 = (num & x000000FF) >> 0 ;
byte1 = (num & x0000FF00) >> 8 ;
byte2 = (num & x00FF0000) >> 16 ;
byte3 = (num & xFF000000) >> 24 ;return((byte0 << 24) | (byte1 << 16) | (byte2 << 8) |
(byte3 << 0));
}
```

12. Program to find if a machine is big endian or little endian?

```
01      #include "stdio.h"
02      #define BIG_ENDIAN 0
03      #define LITTLE_ENDIAN 1
04      int main()
05             {
06                 int value;
07            value = endian();
08            if (value == 1)
09      printf("Machine is little endian\n",value);
10      else
11      printf("Machine is Big Endian\n",value);
12   }
13       int endian() {
14       short int word = 0x0001;
15   char *byte = (char *) &word;
16   return (byte[0] ? LITTLE_ENDIAN : BIG_ENDIAN);
17                                                      }
```

13. Swap 2 variables without using temporary variable!

```
a = a + b
b = a – b
a = a – b
```

14. Write a program to generate the Fibonacci Series?

```
#include<stdio.h>
#include<conio.h>
main()
{
int n,i,c,a=0,b=1;
printf("Enter Fibonacci series of nth term : ");
scanf("%d",&n);
printf("%d %d ",a,b);
for(i=0;i<=(n-3);i++)
```

```
{
c=a+b;
a=b;
b=c;
printf("%d ",c);
}
getch();
}Output :
Enter Fibonacci series of nth term : 7
0 1 1 2 3 5 8
```

15. Write a program to find unique numbers in an array?

Answer:

```
for (i=1;i<=array.length;i++) {
found=false;
for (k=i+1;k<=array.length;k++) {
if (array[i]==array[k]) found=true;
}
if (!found) println(array[i]);
}
```

16. Write a C program to print Equilateral Triangle using numbers?

```
01    /* C program to print Equilateral Triangle*/
02         #include<stdio.h>
03         main()
04            {
05               int i,j,k,n;
06
07    printf("Enter number of rows  of the triangle \n");
08         scanf("%d",&n);
09
10         for(i=1;i<=n;i++)
11         {
12            for(j=1;j<=n-i;j++)
13            {
14                printf(" ");
15            }
16         for(k=1;k<=(2*i)-1;k++)
17            {
18                printf("i");
19            }
20            printf("\n");
21         }
22               getch();
```

17. Write a program for deletion and insertion of a node in single linked list?

```
#include<stdio.h>
#include<stdlib.h>
typedef struct Node
{
int data;
```

```c
struct Node *next;
}node;
void insert(node *pointer, int data)
{
/* Iterate through the list till we encounter the last node.*/
while(pointer->next!=NULL)
{
pointer = pointer -> next;
}
/* Allocate memory for the new node and put data in it.*/
pointer->next = (node *)malloc(sizeof(node));
pointer = pointer->next;
pointer->data = data;
pointer->next = NULL;
}
int find(node *pointer, int key)
{
pointer =  pointer -> next; //First node is dummy node.
/* Iterate through the entire linked list and search for the key. */
while(pointer!=NULL)
{
if(pointer->data == key) //key is found.
{
return 1;
}
pointer = pointer -> next;//Search in the next node.
}
/*Key is not found */
return 0;
}
void delete(node *pointer, int data)
{
/* Go to the node for which the node next to it has to be deleted */
while(pointer->next!=NULL && (pointer->next)->data != data)
{
pointer = pointer -> next;
}
if(pointer->next==NULL)
{
printf("Element %d is not present in the list\n",data);
return;
}
/* Now pointer points to a node and the node next to it has to be removed */
node *temp;
temp = pointer -> next;
/*temp points to the node which has to be removed*/
pointer->next = temp->next;
```

```c
/* We removed the node which is next to the pointer (which is also temp) */
free(temp);
/* Beacuse we deleted the node, we no longer require the memory used for it .
free() will deallocate the memory.
*/
return;
}
void print(node *pointer)
{
if(pointer==NULL)
{
return;
}
printf("%d ",pointer->data);
print(pointer->next);
}
int main()
{
/* start always points to the first node of the linked list.
temp is used to point to the last node of the linked list.*/
node *start,*temp;
start = (node *)malloc(sizeof(node));
temp = start;
temp -> next = NULL;
/* Here in this code, we take the first node as a dummy node.
The first node does not contain data, but it used because to avoid handling special cases
in insert and delete functions.
*/
printf("1. Insert\n");
printf("2. Delete\n");
printf("3. Print\n");
printf("4. Find\n");
while(1)
{
int query;
scanf("%d",&query);
if(query==1)
{
int data;
scanf("%d",&data);
insert(start,data);
}
else if(query==2)
{
int data;
scanf("%d",&data);
delete(start,data);
```

```
}
else if(query==3)
{
printf("The list is ");
print(start->next);
printf("\n");
}
else if(query==4)
{
int data;
scanf("%d",&data);
int status = find(start,data);
if(status)
{
printf("Element Found\n");
}
else
{
printf("Element Not Found\n");                    }
}
}}
```

18. Can a variable be both *const* and *volatile*? Yes. The const modifier means that this code cannot change the value of the variable, but that does not mean that the value cannot be changed by means outside this code. For instance, in the example in FAQ 8, the timer structure was accessed through a volatile const pointer. The function itself did not change the value of the timer, so it was declared const. However, the value was changed by hardware on the computer, so it was declared volatile. If a variable is both const and volatile, the two modifiers can appear in either order.

19. what are Constant and Volatile Qualifiers?

const

- constis used with a datatype declaration or definition to specify an unchanging value
  - Examples:
    - const int five = 5;

      const double pi = 3.141593;

- constobjects may not be changed
  - The following are illegal:
    - const int five = 5;

      const double pi = 3.141593;

      ▪

      pi = 3.2;

five = 6;

volatile
- volatilespecifies a variable whose value may be changed by processes outside the current program
- One example of a volatileobject might be a buffer used to exchange data with an external device:
- int

- check_iobuf(void)

- {

- volatile int iobuf;

- int val;

-

- while (iobuf == 0) {

- }

- val = iobuf;

- iobuf = 0;

- return(val);

}

if iobuf had not been declared volatile, the compiler would notice that nothing happens inside the loop and thus eliminate the loop
- const and volatile can be used together
   - An input-only buffer for an external device could be declared as const volatile (or volatile const, order is not important) to make sure the compiler knows that the variable should not be changed (because it is input-only) and that its value may be altered by processes other than the current program

The keywords const and volatile can be applied to any declaration, including those of structures, unions, enumerated types or typedef names. Applying them to a declaration is called qualifying the declaration—that's why const and volatile are called type qualifiers, rather than type specifiers. Here are a few representative examples:
volatile i;

```
volatile int j;

const long q;

const volatile unsigned long int rt_clk;

struct{

    const long int li;

    signed char sc;

}volatile vs;
```

Don't be put off; some of them are deliberately complicated: what they mean will be explained later. Remember that they could also be further complicated by introducing storage class specifications as well! In fact, the truly spectacular

```
extern const volatile unsigned long int rt_clk;
```

is a strong possibility in some real-time operating system kernels.

Const

Let's look at what is meant when const is used. It's really quite simple: const means that something is not modifiable, so a data object that is declared with const as a part of its type specification must not be assigned to in any way during the run of a program. It is very likely that the definition of the object will contain an initializer (otherwise, since you can't assign to it, how would it ever get a value?), but this is not always the case. For example, if you were accessing a hardware port at a fixed memory address and promised only to read from it, then it would be declared to be const but not initialized.
Taking the address of a data object of a type which isn't const and putting it into a pointer to the const-qualified version of the same type is both safe and explicitly permitted; you will be able to use the pointer to inspect the object, but not modify it. Putting the address of a const type into a pointer to the unqualified type is much more dangerous and consequently prohibited (although you can get around this by using a cast). Here is an example:
```
#include <stdio.h>

#include <stdlib.h>



main(){
```

```
int i;

const int ci = 123;



/* declare a pointer to a const.. */

const int *cpi;



/* ordinary pointer to a non-const */

int *ncpi;



cpi = &ci;

ncpi = &i;



/*

 * this is allowed

 */

cpi = ncpi;



/*

 * this needs a cast

 * because it is usually a big mistake,

 * see what it permits below.
```

```
    */

    ncpi = (int *)cpi;



    /*

    * now to get undefined behaviour...

    * modify a const through a pointer

    */

    *ncpi = 0;



    exit(EXIT_SUCCESS);

}
```

Example 8.3
As the example shows, it is possible to take the address of a constant object, generate a
pointer to a non-constant, then use the new pointer. This is an *error* in your program and
results in undefined behaviour.
The main intention of introducing const objects was to allow them to be put into read-only
store, and to permit compilers to do extra consistency checking in a program. Unless you
defeat the intent by doing naughty things with pointers, a compiler is able to check
that const objects are not modified explicitly by the user.
An interesting extra feature pops up now. What does this mean?

```
char c;

char *const cp = &c;
```

It's simple really; cp is a pointer to a char, which is exactly what it would be if
the const weren't there. The const means that cp is not to be modified, although whatever it
points to can be—the pointer is constant, not the thing that it points to. The other way round
is
```
const char *cp;
```

which means that now cp is an ordinary, modifiable pointer, but the thing that it points to must not be modified. So, depending on what you choose to do, both the pointer and the thing it points to may be modifiable or not; just choose the appropriate declaration.

Volatile

After const, we treat volatile. The reason for having this type qualifier is mainly to do with the problems that are encountered in real-time or embedded systems programming using C. Imagine that you are writing code that controls a hardware device by placing appropriate values in hardware registers at known absolute addresses.

Let's imagine that the device has two registers, each 16 bits long, at ascending memory addresses; the first one is the control and status register (csr) and the second is a data port. The traditional way of accessing such a device is like this:

```
/* Standard C example but without const or volatile */

/*

* Declare the device registers

* Whether to use int or short

* is implementation dependent

*/



struct devregs{

    unsigned short  csr;    /* control & status */

    unsigned short  data;   /* data port */

};



/* bit patterns in the csr */

#define ERROR   0x1

#define READY   0x2
```

```c
#define RESET   0x4


/* absolute address of the device */

#define DEVADDR ((struct devregs *)0xffff0004)


/* number of such devices in system */

#define NDEVS   4


/*

* Busy-wait function to read a byte from device n.

* check range of device number.

* Wait until READY or ERROR

* if no error, read byte, return it

* otherwise reset error, return 0xffff

*/

unsigned int read_dev(unsigned devno){


    struct devregs *dvp = DEVADDR + devno;


    if(devno >= NDEVS)

        return(0xffff);
```

```c
        while((dvp->csr & (READY | ERROR)) == 0)

                ; /* NULL - wait till done */



        if(dvp->csr & ERROR){

                dvp->csr = RESET;

                return(0xffff);

        }



        return((dvp->data) & 0xff);

}
```

Example 8.4
The technique of using a structure declaration to describe the device register layout and names is very common practice. Notice that there aren't actually any objects of that type defined, so the declaration simply indicates the structure without using up any store.

To access the device registers, an appropriately cast constant is used as if it were pointing to such a structure, but of course it points to memory addresses instead.

However, a major problem with previous C compilers would be in the while loop which tests the status register and waits for the ERROR or READY bit to come on. Any self-respecting optimizing compiler would notice that the loop tests the same memory address over and over again. It would almost certainly arrange to reference memory once only, and copy the value into a hardware register, thus speeding up the loop. This is, of course, exactly what we don't want: this is one of the few places where we must look at the place where the pointer points, every time around the loop.
Because of this problem, most C compilers have been unable to make that sort of optimization in the past. To remove the problem (and other similar ones to do with when to write to where a pointer points), the keyword volatile was introduced. It tells the compiler that the object is subject to sudden change for reasons which cannot be predicted from a study of the program itself, and forces every reference to such an object to be a genuine reference. Here is how you would rewrite the example, making use of const and volatile to get what you want.

```c
/*
 * Declare the device registers

 * Whether to use int or short

 * is implementation dependent
 */


struct devregs{

    unsigned short volatile csr;

    unsigned short const volatile data;

};


/* bit patterns in the csr */

#define ERROR   0x1

#define READY   0x2

#define RESET   0x4


/* absolute address of the device */

#define DEVADDR ((struct devregs *)0xffff0004)


/* number of such devices in system */

#define NDEVS   4
```

```c
/*

 * Busy-wait function to read a byte from device n.

 * check range of device number.

 * Wait until READY or ERROR

 * if no error, read byte, return it

 * otherwise reset error, return 0xffff

 */

unsigned int read_dev(unsigned devno){


    struct devregs * const dvp = DEVADDR + devno;


    if(devno >= NDEVS)

        return(0xffff);


    while((dvp->csr & (READY | ERROR)) == 0)

        ; /* NULL - wait till done */


    if(dvp->csr & ERROR){

        dvp->csr = RESET;

        return(0xffff);
```

```
      }


      return((dvp->data) & 0xff);


}
```

Example 8.5
The rules about mixing volatile and regular types resemble those for const. A pointer to
a volatile object can be assigned the address of a regular object with safety, but it is
dangerous (and needs a cast) to take the address of a volatile object and put it into a pointer to
a regular object. Using such a derived pointer results in undefined behaviour.
If an array, union or structure is declared with const or volatile attributes, then all of the
members take on that attribute too. This makes sense when you think about it—how could a
member of a const structure be modifiable?
That means that an alternative rewrite of the last example would be possible. Instead of
declaring the device registers to be volatile in the structure, the pointer could have been
declared to point to a volatile structure instead, like this:

```
struct devregs{

     unsigned short  csr;   /* control & status */

     unsigned short  data;   /* data port */

};

volatile struct devregs *const dvp=DEVADDR+devno;
```

Since dvp points to a volatile object, it not permitted to optimize references through the
pointer. Our feeling is that, although this would work, it is bad style. The volatile declaration
belongs in the structure: it is the device registers which are volatile and that is where the
information should be kept; it reinforces the fact for a human reader.
So, for any object likely to be subject to modification either by hardware or asynchronous
interrupt service routines, the volatile type qualifier is important.

Now, just when you thought that you understood all that, here comes the final twist. A
declaration like this:

```
volatile struct devregs{

    /* stuff */

}v_decl;
```

declares the type struct devregs and also a volatile-qualified object of that type, called v_decl. A later declaration like this
struct devregs nv_decl;

declares nv_decl which is *not* qualified with volatile! The qualification is *not* part of the type of struct devregs but applies only to the declaration of v_decl. Look at it this way round, which perhaps makes the situation more clear (the two declarations are the same in their effect):
struct devregs{

    /* stuff */

}volatile v_decl;

If you do want to get a shorthand way of attaching a qualifier to another type, you can use typedef to do it:
struct x{

    int a;

};

typedef const struct x csx;

csx const_sx;

struct x non_const_sx = {1};

const_sx = non_const_sx;   /* error - attempt to modify a const */

20.What are the differences between a union and a structure in C?
A union is a way of providing an alternate way of describing the same memory area. In this way, you could have a struct that contains a union, so that the "static", or similar portion of the data is described first, and the portion that changes is described by the union. The idea of a union could be handled in a different way by having 2 different structs defined, and making a pointer to each kind of struct. The pointer to struct "a" could be assigned to the value of a buffer, and the pointer to struct "b" could be assigned to the same buffer, but now a->some field and b->some otherfield are both located in the same buffer. That is the idea behind a union. It gives different ways to break down the same buffer area.
The difference between structure and union are: 1. union allocates the memory equal to the

maximum memory required by the member of the union but structure allocates the memory equal to the total memory required by the members. 2. In union, one block is used by all the member of the union but in case of structure, each member have their own memory space

Difference Between Stucture and Union :

Structure	Union
i. Access Members	
We can access all the members of structure at anytime.	Only one member of union can be accessed at anytime.
ii. Memory Allocation	
Memory is allocated for all variables.	Allocates memory for variable which variable require more memory.
iii. Initialization	
All members of structure can be initialized	Only the first member of a union can be initialized.
iv. Keyword	
'struct' keyword is used to declare structure.	'union' keyword is used to declare union.
v. Syntax	

```
struct struct_name                    union union_name

{                                     {

    structure element 1;                  union element 1;

    structure element 2;                  union element 2;

        ----------                            ----------


        ----------                            ----------

    structure element n;                  union element n;

}struct_var_nm;                        }union_var_nm;
```

vi. Example

```
struct item_mst                       union item_mst

{                                     {

    int rno;                              int rno;

    char nm[50];                          char nm[50];
```

}it;   }it;

Difference in their Usage:

While structure enables us treat a number of different variables stored at different in memory, a union enables us to treat the same space in memory as a number of different variables. That is a Union offers a way for a section of memory to be treated as a variable of one type on one occasion and as a different variable of a different type on another occasion.

There is frequent requirement while interacting with hardware to access access a byte or group of bytes simultaneously and sometimes each byte individually. Usually union is the answer.

========Difference With example========

Lets say a structure containing an int,char and float is created and a union containing int char float are declared.

struct TT{ int a; float b; char c; } Union UU{ int a; float b; char c; }
sizeof TT(struct) would be >9 bytes (compiler dependent-if int,float, char are taken as 4,4,1)
sizeof UU(Union) would be 4 bytes as supposed from above.If a variable in double exists in union then the size of union and struct would be 8 bytes and cumulative size of all variables in struct.

Detailed Example:

struct foo
{
char c;
long l;
char *p;
};

union bar
{
char c;
long l;
char *p;
};

A struct foo contains all of the elements c, l, and p. Each element is separate and distinct.

A union bar contains only one of the elements c, l, and p at any given time. Each element is stored in the same memory location (well, they all
start at the same memory location), and you can only refer to the element which was last stored. (ie: after "barptr->c = 2;" you cannot reference
any of the other elements, such as "barptr->p" without invoking undefined behavior.)

Try the following program. (Yes, I know it invokes the above-mentioned "undefined behavior", but most likely will give some sort of output on most computers.)

```
==========
#include

struct foo
{
char c;
long l;
char *p;
};

union bar
{
char c;
long l;
char *p;
};

int main(int argc,char *argv[])
{
struct foo myfoo;
union bar mybar;

myfoo.c = 1;
myfoo.l = 2L;
myfoo.p = "This is myfoo";

mybar.c = 1;
mybar.l = 2L;
mybar.p = "This is mybar";

printf("myfoo: %d %ld %s\n",myfoo.c,myfoo.l,myfoo.p);
printf("mybar: %d %ld %s\n",mybar.c,mybar.l,mybar.p);

return 0;
}

==========
```

On my system, I get:

```
myfoo: 1 2 This is myfoo
mybar: 100 4197476 This is mybar
==========
```

credit to original author.

Structure: Structure is a combination elements, which can be predefined data types or other structure. The length/size of the structure is the sum of the length of its elements.

In C, structures cannot contain functions. in C++ it can.

Union: Union is a combination elements, which can be predefined data types or other union . But, the size/length of union is the maximum of internal elements.

the *sizeof()* operator returns the size slightly more than calculated size due to padding, which again depends on OS

== Answer == Union allocates the memory equal to the maximum memory required by the member of the union but structure allocates the memory equal to the total memory required by the members. In union,one block is used by all the member of the union but in case of structure, each member have their own memory space.

21. What is meant by structure padding?

Answer: compilers pad structures to optimize data transfers. This is an hardware architecture issue. Most modern CPUs perform best when fundamental types, like 'int' or 'float', are aligned on memory boundaries of a particular size (eg. often a 4byte word on 32bit archs). Many architectures don't allow misaligned access or if they do inoccur a performance penality. When a compiler processes a structure declaration it will add extra bytes between fields to meet alignment needs.

Most processors require specific memory alignment on variables certain types. Normally the minimum alignment is the size of the basic type in question, for instance this is common

char variables can be byte aligned and appear at any byte boundary

short (2 byte) variables must be 2 byte aligned, they can appear at any even byte boundary. This means that 0x10004567 is not a valid location for a short variable but 0x10004566 is.

long (4 byte) variables must be 4 byte aligned, they can only appear at byte boundaries that are a multiple of 4 bytes. This means that 0x10004566 is not a valid location for a long variable but 0x10004568 is.

Structure padding occurs because the members of the structure must appear at the correct byte boundary, to achieve this the compiler puts in padding bytes (or bits if bit fields are in use) so that the structure members appear in the correct location. Additionally the size of the structure must be such that in an array of the structures all the structures are correctly aligned in memory so there may be padding bytes at the end of the structure too

```
struct example {
char c1;
short s1;
char c2;
long l1;
char c3;
}
```

In this structure, assuming the alignment scheme I have previously stated then c1 can appear at any byte boundary, however s1 must appear at a 2 byte boundary so there is a padding byte between c1 and s1.

c2 can then appear in the available memory location, however l1 must be at a 4 byte boundary so there are 3 padding bytes between c2 and l1

c3 then appear in the available memory location, however because the structure contains a long member the structure must be 4 byte aligned and must be a multiple of 4 bytes in size. Therefore there are 3 padding bytes at the end of the structure. It would appear in memory in this order

c1
padding byte
s1 byte 1
s1 byte 2
c2
padding byte
padding byte
padding byte
l1 byte 1
l1 byte 2
l1 byte 3
l1 byte 4
c3
padding byte
padding byte
padding byte

The structure would be 16 bytes long.

re-written like this

```
struct example {
long l1;
short s1;
char c1;
char c2;
char c3;
}
```

Then l1 appears at the correct byte alignment. s1 will be correctly aligned so no need for padding between l1 and s1. c1, c2, c3 can appear at any location. The structure must be a multiple of 4 bytes in size since it contains a long so 3 padding bytes appear after c3
It appears in memory in the order

l1 byte 1
l1 byte 2
l1 byte 3
l1 byte 4
s1 byte 1
s1 byte 2
c1
c2
c3
padding byte
padding byte
padding byte

and is only 12 bytes long.

I should point out that structure packing is platform and compiler (and in some cases compiler switch) dependent.

Memory Pools are just a section of memory reserved for allocating temporarily to other parts of the application

A memory leak occurs when you allocate some memory from the heap (or a pool) and then delete all references to that memory without returning it to the pool it was allocated from.

Program:

```
struct  MyStructA {
    char a;
    char b;
    int   c;
};

struct MyStructB {
    char a;
    int   c;
    char b;
};

int main(void) {
    int sizeA = sizeof(struct MyStructA);
    int sizeB = sizeof(struct MyStructB);

    printf("A = %d\n", sizeA);
    printf("B = %d\n", sizeB);

    return 0;
}
```

22. What is the difference between macro and constant variables in C?

Macros are replaced by preprocessor, but in constant data type will be checked by compiler. Macros are replaced without checking the values sometimes the programmer want to change values only in a single function at that prefer to use constant than a macro.

The first technique comes from the C programming language. Constants may be defined using the preprocessor directive, #define The preprocessor is a program that modifies your source file prior to compilation. Common preprocessor directives are #include, which is used to include additional code into your source file, #define, which is used to define a constant and #if/#endif, which can be used to conditionally determine which parts of your code will be compiled. The #define directive is used as follows.

#define pi 3.1415#define id_no 12345

Wherever the constant appears in your source file, the preprocessor replaces it by its value. So, for instance, every "pi" in your source code will be replace by 3.1415. The compiler will only see the value 3.1415 in your code, not "pi". The problem with this technique is that the replacement is done lexically, without any type checking, without any bound checking and without any scope checking. Every "pi" is just replaced by its value. The technique is outdated, exists to support legacy code and should be avoided.

Const

The second technique is to use the keyword const when defining a variable. When used the compiler will catch attempts to modify variables that have been declared const.

const float pi = 3.1415;const int id_no = 12345;

There are two main advantages over the first technique.

First, the type of the constant is defined. "pi" is float. "id_no" is int. This allows some type checking by the compiler.&nbs p;
Second, these constants are variables with a definite scope. The scope of a variable relates to parts of your program in which it is defined. Some variables may exist only in certain functions or in certain blocks of code.

Ex : You may want to use "id_no" in one function
and a completely unrelated "id_no" in your main program.

23. What is difference between re-entrant function and recursive function in C?

Answer: Re entrant function is a function which guaranteed that which can be work well under multi threaded environment. mean while function is access by one thread, another thread can call it mean there is separate execution stack and handling for each. So function should not contain any static or shared variable which can harm or disturb the execution. Mean function which can be called by thread, while running from another thread safely and properly

Example:

int t;

```c
void swap(int *x, int *y)

{

int s;



s = t; // save global variable

t = *x;

*x = *y;

// hardware interrupt might invoke isr() here!

*y = t;

t = s; // restore global variable

}



void isr()

{

int x = 1, y = 2;

swap(&x, &y);

}
```

Recursive function Example:
```c
void doll ( int size )
{
if ( size == 0 )  // No doll can be smaller than 1 atom (10^0==1) so doesn't call itself
return;        // Return does not have to return something, it can be used
//  to exit a function
```

```
doll ( size – 1 ); // Decrements the size variable so the next doll will be smaller.
}
int main()
{
doll ( 10 ); //Starts off with a large doll (it's a logarithmic scale)
```

24. What is V-Model ? What are the benefits?

The **V-model** represents a software development process (also applicable to hardware development) which may be considered an extension of the waterfall model. Instead of moving down in a linear way, the process steps are bent upwards after the coding phase, to form the typical V shape. The V-Model demonstrates the relationships between each phase of the development life cycle and its associated phase of testing.
The V model has a number of benefits:

1. Systems development projects usually have a test approach, or test strategy document, which defines how testing will be performed throughout the lifecycle of the project. The V model provides a consistent basis and standard for part of that strategy.

2. The V model explicitly suggests that testing (quality assurance) should be considered early on in the life of a project. Testing and fixing can be done at any stage in the lifecycle. However, the cost of finding and fixing faults increases dramatically as development progresses. Evidence suggests that if a fault uncovered during design costs 1.0 monetary unit to correct, then the same fault uncovered just before testing will cost 6.5 units, during testing 15 units, and after release between 60 and 100 units. The need to find faults as soon as possible reinforces the need for the quality assurance of documents such as the requirements specification and the functional specification. This is performed using static testing techniques such as inspections and walkthroughs.

3. It introduces the idea of specifying test requirements and expected outcomes prior to performing the actual tests. For example, the acceptance tests are performed against a specification of requirements, rather than against some criteria dreamed up when the acceptance stage has been reached

4. The V model provides a focus for defining the testing that must take place within each stage. The definition of testing is assisted by the idea of entry and exit criteria. Hence, the model can be used to define the state a deliverable must be in before it can enter and leave each stage. The exit criteria of one stage are usually the entry criteria of the next. In many organizations, there is concern about the quality of the program code released by individual programmers. Some programmers release code that appears to be fault-free, while others release code that still has many faults in it. The problem of programmers releasing code with different levels of robustness would be addressed in the exit criteria of unit design and unit testing. Unit design would require programmers to specify their intended test cases before they wrote any program code. Coding could not begin until these test cases had been agreed with an appropriate manager. Second, the test cases would have to be conducted successfully before the program could leave the unit test stage and be released to integration testing.

5. Finally, the V model provides a basis for defining who is responsible for performing the testing at each stage. Here are some typical responsibilities:

- acceptance testing performed by users
- system testing performed by system testers
- integration testing performed by program team leaders
- unit testing performed by programmers.

The V model is therefore an excellent basis for the partitioning of testing, highlighting the fact that all the participants in the development of a system have a responsibility for quality assurance and testing.

25. What is meant by Black box testing and white box testing?
White-box testing (also known as **clear box testing, glass box testing, transparent box testing**, and **structural testing**) is a method of testing software that tests internal structures or workings of an application, as opposed to its functionality (i.e. black-box testing). In white-box testing an internal perspective of the system, as well as programming skills, are used to design test cases. The tester chooses inputs to exercise paths through the code and determine the appropriate outputs. This is analogous to testing nodes in a circuit, e.g. in-circuit testing (ICT).

While white-box testing can be applied at the unit, integration and system levels of the software testing process, it is usually done at the unit level. It can test paths within a unit, paths between units during integration, and between subsystems during a system–level test. Though this method of test design can uncover many errors or problems, it might not detect unimplemented parts of the specification or missing requirement.

Black-box testing is a method of software testing that tests the functionality of an application as opposed to its internal structures or workings (see white-box testing). This method of test can be applied to all levels of software testing: unit, integration, system and acceptance. It typically comprises most if not all testing at higher levels, but can also dominate unit testing as well.

White Box Testing: Means testing the application with coding /programming knowledge. That means the tester has to correct the code also.
Black box testing: Testing the application without coding /programming knowledge that means the tester doesn't require coding knowledge. Just he examines the application external functional behaviour and GUI features.

Sl.No	Black Box	White Box
1	Focuses on the functionality of the system Techniques used are : - Equivalence partitioning - Boundary-value analysis - Error guessing - Race conditions - Cause-effect graphing	Focuses on the structure (Program) of the system Techniques used are: - Basis Path Testing - Flow Graph Notation - Control Structure Testing 1. Condition Testing
2	- Syntax testing	

- State transition testing
- Graph matrix

2. Data Flow testing

- Loop Testing
1. Simple Loops

2. Nested Loops

3. Concatenated Loops

4. Unstructured Loops

3	Tester can be non technical	Tester should be technical
4	Helps to identify the vagueness and contradiction in functional specifications	Helps to identify the logical and coding issues.

26. What are the types of testings?

unit testing

Component testing

Integration testing

System testing

27. The Difference between Bit Rate and Baud Rate?

The difference between Bit and Baud rate is complicated and intertwining. Both are dependent and inter-related. But the simplest explanation is that a Bit Rate is how many data bits are transmitted per second. A baud Rate is the number of times per second a signal in a communications channel changes.

Bit rates measure the number of data bits (that is 0's and 1's) transmitted in one second in a communication channel. A figure of 2400 bits per second means 2400 zeros or ones can be transmitted in one second, hence the abbreviation "bps." Individual characters (for example letters or numbers) that are also referred to as bytes are composed of several bits.

A baud rate is the number of times a signal in a communications channel changes state or varies. For example, a 2400 baud rate means that the channel can change states up to 2400 times per second. The term "change state" means that it can change from 0 to 1 or from 1 to 0 up to X (in this case, 2400) times per second. It also refers to the actual state of the connection, such as voltage, frequency, or phase level).

The main difference between the two is that one change of state can transmit one bit, or slightly more or less than one bit, that depends on the modulation technique used. So the bit rate (bps) and baud rate (baud per second) have this connection:

bps = baud per second x the number of bit per baud

The modulation technique determines the number of bit per baud. Here are two examples:

When FSK (Frequency Shift Keying, a transmission technique) is used, each baud transmits one bit. Only one change in state is required to send a bit. Thus, the modem's bps rate is equal to the baud rate. When a baud rate of 2400 is used, a modulation technique called phase modulation that transmits four bits per baud is used. So:

2400 baud x 4 bits per baud = 9600 bps

Such modems are capable of 9600 bps operation.

3)Difference between flash and EEprom
  1. Similarities
  - Both flash and EEPROM are digital storage methods used by computers and other devices. Both are non-volatile ROM technologies to which you can write and from which you can erase multiple times.

Differences

  - The primary difference between flash and EEPROM is the way they erase data. While EEPROM destroys the individual bytes of memory used to store data, flash devices can only erase memory in larger blocks. This makes flash devices faster at rewriting, as they can affect large portions of memory at once. Since a rewrite may affect unused blocks of data, it also adds unnecessarily to usage of the device, shortening its lifespan in comparison with EEPROM.

Usage

  - Flash storage is commonly used in USB memory drives and solid state hard drives. EEPROM is used in a variety of devices, from programmable VCRs to CD players.

28. Can structures be passed to the functions by value?
Ans: yes structures can be passed by value. But unnecessary memory wastage.

29. Why cannot arrays be passed by values to functions?
Ans : When a array is passed to a function, the array is internally changed to a µpointer. And pointers are always passed by reference.

30. What is meant by static functions?
Ans: static functions are functions that are only visible to other functions in the same file
Example:

main.c
```
#include <STDIO.H>




main()
```

```
    {

    Func1();



    Func2();

    }
```

funcs.c
```
/**********************************

  *

  * Function declarations (prototypes).

  *

  **********************************/



/* Func1 is only visable to functions in this file. */



static void Func1(void);

/* Func2 is visable to all functions. */

void Func2(void);

/**********************************
 *
 * Function definitions
 *
 **********************************/

void Func1(void)
{
  puts("Func1 called");
}
```

```
/***********************************/   void Func2(void)
{
  puts("Func2 called");
}
```

31. Difference between declaration, definition & initialization?

Ans: A declaration introduces a name – an identifier – to the compiler. It tells the compiler "This function or this variable exists somewhere, and here is what it should look like."
A definition, on the other hand, says: "Make this variable here" or "Make this function here." It allocates storage for the name. This meaning works whether
you're talking about a variable or a function; in either case, at the point of definition the compiler allocates storage.

extern const int x = 1; /* Initialization */
This initialization establishes this as a definition, not a declaration.

extern const int x; /* Declaration */
This declaration in C++ means that the definition exists elsewhere.

32. What is the difference between pass by value by reference in c and pass by reference in c?

Pass By Reference :
In Pass by reference address of the variable is passed to a function. Whatever changes made to the formal parameter will affect to the actual parameters
– Same memory location is used for both variables.(Formal and Actual)-
– it is useful when you required to return more then 1 values

Pass By Value:
– In this method value of the variable is passed. Changes made to formal will not affect the actual parameters.
– Different memory locations will be created for both variables.
– Here there will be temporary variable created in the function stack which does not affect the original variable.

33. What is the difference between flash memory, EPROM and EEPROM?

EEPROM is an older, more reliable technology. It is somewhat slower than Flash. Flash and EEPROM are very similar, but there is a subtle difference. Flash and EEPROM both use quantum cells to trap electons. Each cell represents one bit of data. The presence – or absence – of electons in a cell indicates whether the bit is a 1 or 0.The cells have a finite life – every time a cell is erased, it wears out a little bit. In EEPROM, cells are erased one-by-one. The only cells erased are those which are 1 but need to be zero. (Writing a 1 to a cell that's 0 causes very little wear, IIRC)In Flash, a large block is erased all at once. In some devices, this "block" is the entire device. So in flash, cells are erased whether they need it or not. This cuts down on the lifespan of the device, but is much, much faster than the EEPROM method of going cell-by-cell.

Erasure method: Both Flash and EEPROM erase cells by means of an electric field. I think it is high-frequency and "pops" the electrons out of the Other similar devices are EPROM (sometimes UVEPROM) and OTPROM (sometimes PROM). EPROM/UVEPROM lacks the structures that generate the electrical field for erasure. These devices have a window on top, usually covered by a paper sticker. To erase, the sticker is removed and the device is exposed to intense ultraviolet light for 30-45 minutes. The only difference between OTPROM and UVEPROM is that OTPROM lacks the UV window – there is no way to erase the data. Adding the UV window to the device package significantly increases cost, so there is a niche for one-time-programmable devices

Erasable Programmable Read Only Memory Chips

The information stored in an EPROM chip can be **erased** by exposing the chip to strong UV light. EPROM chips are easily recognized by the small quartz window used for erasure. Once erased the chip can be **re-programmed**.

EPROM is more expensive to buy per unit cost, but can prove cheaper in the long run for some applications. For example if PROM was used for firmware that needed upgraded every 6 months or so – it could prove quite expensive buying new chips!

Electronically Erasable Programmable Read Only Memory

This has the added advantage that the information stored can be re-written in **blocks** and hence can be used to store **system settings** that the user may want to change periodically. This solid state memory has considerably reduced in price over recent years and is nowadays commonly used to store system settings such as BIOS settings

34. What is difference between Volatile & Non Volatile Memory?

Volatile memory

Volatile memory is computer memory that requires power to maintain the stored information. Most modern semiconductor volatile memory is either Static RAM (see SRAM) or dynamic RAM (see DRAM). SRAM retains its contents as long as the power is connected and is easy to interface to but uses six transistors per bit. Dynamic RAM is more complicated to interface to and control and needs regular refresh cycles to prevent its contents being lost. However, DRAM uses only one transistor and a capacitor per bit, allowing it to reach much higher densities and, with more bits on a memory chip, be much cheaper per bit. SRAM is not worthwhile for desktop system memory, where DRAM dominates, but is used for their cache memories. SRAM is commonplace in small embedded systems, which might only need tens of kilobytes or less. Forthcoming volatile memory technologies that hope to replace or compete with SRAM and DRAM include Z-RAM, TTRAM, A-RAM and ETA RAM.

Non-volatile memory

Non-volatile memory is computer memory that can retain the stored information even when not powered. Examples of non-volatile memory include read-only memory (see ROM), flash memory, most types of magnetic computer storage devices (e.g. hard disks, floppy discs and magnetic tape), optical discs, and early computer storage methods such as paper tape and punched cards. Forthcoming non-volatile memory technologies include FeRAM, CBRAM, PRAM, SONOS, RRAM, Racetrack memory, NRAM and Millipede.

35. What is a reentrant function?

A **reentrant function** is a function which can be safely executed **concurrently**. This means it should allow a re-entry while it is running. The reentrant function should work only on the data given by the calling function. It must not have any static data also.

The term "reentrant" is used to refer to side wall profiles of the nozzles, wherein exit diameters of the nozzles are smaller than entrance diameters of the nozzles so that the side walls of the nozzles are not perpendicular to a plane defined by an exit surface of the nozzle member.

A Reentrant function is a function which guaranteed that which can be work well under multi threaded environment. Mean while function is access by one thread, another thread can call it… mean there is separate execution stack and handling for each. So function should not contain any static or shared variable which can harm or disturb the execution.

36. Why we are using UDS, If CAN support diagnostics communication?

In CAN it will support internal diagnostic messages. UDS & KWP2000 are used for to test with external tester and to know the type of problem

37. How to recover from CAN- Busoff?

To distinguish between temporary and permanent failures every CAN bus controller has two Error Counters: The REC (Receive Error Counter) and the TEC (Transmit Error Counter). The counters are incremented upon detected errors respectively are decremented upon correct transmissions or receptions. Depending on the counter values the state of the node is changed: The initial state of a CAN bus controller is Error Active that means the controller can send active Error Flags. The controller gets in the Error Passive state if there is an accumulation of errors.

On CAN bus controller failure or an extreme accumulation of errors there is a state transition to Bus Off. The controller is disconnected from the bus by setting it in a state of high-resistance. **The Bus Off state should only be left by a software reset. After software reset the CAN bus controller has to wait for 128 x 11 recessive bits to transmit a frame. This is because other nodes may pending transmission requests. It is recommended not to start an hardware reset because the wait time rule will not be followed then.**

37. What is Virtual functional bus?

Virtual function bus can be described as a system modeling and communication concept. It is logical entity that facilitates the
concept of relocatability within the AUTOSAR software architecture by providing a virtual

infrastructure that is independent from any actual underlying infrastructure and provides all services required for a virtual interaction between AUTOSAR components.

38. Intra and Inter ECU communication?

Intra-ECU which denotes the communication between two software components residing on the same ECU and Inter-ECU which denotes the situation when two software components reside on different ECU's that are connected via a bus network.

39. What is the difference between global and static global variables?

Global variables are variables defined outside of any function. Their scope starts at the point where they are defined and lasts to the end of the file. They have external linkage, which means that in other source files, the same name refers to the same location in memory.
Static global variables are private to the source file where they are defined and do not conflict with other variables in other source files which would have the same name.

40. How to access a Global variables in other files?

Variables declared outside of a block are called **global variables**. Global variables have **program scope**, which means they can be accessed everywhere in the program, and they are only destroyed when the program ends.
Here is an example of declaring a global variable:

```
1   int g_nX; // global variable
2
3   int main()
4   {
5     int nY; // local variable nY
6
7     // global vars can be seen everywhere in program
8     // so we can change their values here
9     g_nX = 5;
10  } // nY is destroyed here
```

Because global variables have program scope, they can be used across multiple files. In the section on programs with multiple files, you learned that in order to use a function declared in another file, you have to use a forward declaration, or a header file.

Similarly, in order to use a global variable that has been declared in another file, you have to use a forward declaration or a header file, along with the **extern** keyword. Extern tells the compiler that you are not declaring a new variable, but instead referring to a variable declared elsewhere.
Here is an example of using a forward declaration style extern:

global.cpp:

```
1   // declaration of g_nValue
2   int g_nValue = 5;
```
main.cpp:

```
1 // extern tells the compiler this variable is declared elsewhere
2 extern int g_nValue;
```

```
3
4 int main()
5 {
6    g_nValue = 7;
7    return 0;
8 }
```

Here is an example of using a header file extern:

global.cpp:

```
1    // declaration of g_nValue
2    int g_nValue = 5;
```

global.h:

```
1 #ifndef GLOBAL_H // header guards
2 #define GLOBAL_H
3
4 // extern tells the compiler this variable is declared elsewhere
5 extern int g_nValue;
6
7 #endif
```

main.cpp:

```
1    #include "global.h"
2    int main()
3    {
4        g_nValue = 7;
5        return 0;
6    }
```

Generally speaking, if a global variable is going to be used in more than 2 files, it's better to use the header file approach. Some programmers place all of a programs global variables in a file calledglobals.cpp, and create a header file named globals.h to be included by other .cpp files that need to use them.

Local variables with the same name as a global variable hide the global variable inside that block. However, the global scope operator (::) can be used to tell the compiler you mean the global version:

```
1 int nValue = 5;
2
3 int main()
4 {
5    int nValue = 7; // hides the global nValue variable
6    nValue++; // increments local nValue, not global nValue
7    ::nValue--; // decrements global nValue, not local nValue
8    return 0;
9 } // local nValue is destroyed
```

However, having local variables with the same name as global variables is usually a recipe for trouble, and should be avoided whenever possible. Using Hungarian Notation, it is common to declare global variables with a "g_" prefix. This is an easy way to differentiate global variable from local variables, and avoid variables being hidden due to naming collisions.

New programmers are often tempted to use lots of global variables, because they are easy to work with, especially when many functions are involved. However, this is a very bad idea. In fact, global variables should generally be avoided completely!

Why global variables are evil

Global variables should be avoided for several reasons, but the primary reason is because they increase your program's complexity immensely. For example, say you were examining a program and you wanted to know what a variable named g_nValue was used for. Because g_nValue is a global, and globals can be used anywhere in the entire program, you'd have to examine every single line of every single file! In a computer program with hundreds of files and millions of lines of code, you can imagine how long this would take!

Second, global variables are dangerous because their values can be changed by any function that is called, and there is no easy way for the programmer to know that this will happen. Consider the following program:

```
1    // declare global variable
2    int g_nMode = 1;
3
4    void doSomething()
5    {
6        g_nMode = 2;
7    }
8
9    int main()
10   {
11       g_nMode = 1;
12
13       doSomething();
14
15       // Programmer expects g_nMode to be 1
16       // But doSomething changed it to 2!
17
18       if (g_nMode == 1)
19           cout << "No threat detected." << endl;
20       else
21           cout << "Launching nuclear missiles..." << endl;
22
23       return 0;
24   }
```

Note that the programmer set g_nMode to 1, and then called doSomething(). Unless the programmer had explicit knowledge that doSomething() was going to change the value of g_nMode, he or she was probably not expecting doSomething() to change the value! Consequently, the rest of main() doesn't work like the programmer expects (and the world is obliterated).

Global variables make every function call potentially dangerous, and the programmer has no easy way of knowing which ones are dangerous and which ones aren't! Local variables are much safer because other functions can not affect them directly. Consequently, global variables should not be used unless there is a very good reason!

41. What is the use of Complex Device Drivers in AUTOSAR?

Since the AUTOSAR layered software architecture restricts direct access to hardware from upper layers, an additional concept is provided in order to bypass that restriction for resource critical and/or Non-AUTOSAR compliant software components. The Complex Device Driver provides an AUTOSAR Interface to the application layer and has direct access to values on the physical layer. This is usually used for the implementation of complex sensor or actuator drivers that need direct control over the underlying hardware.

42. How to set, clear, toggle and checking a single bit in C?

Use the bitwise OR operator (|) to set a bit.

```
number |= 1 << x;
```

That will set bit x.

Clearing a bit

Use the bitwise AND operator (&) to clear a bit.

```
number &= ~(1 << x);
```

That will clear bit x. You must invert the bit string with the bitwise NOT operator (~), then AND it.

Toggling a bit

The XOR operator (^) can be used to toggle a bit.

```
number ^= 1 << x;
```

That will toggle bit x.

Checking a bit

You didn't ask for this but I might as well add it.

To check a bit, AND it with the bit you want to check:

```
bit = number & (1 << x);
```
That will put the value of bit x into the variable bit.

42. What is Watchdog timer?

A watchdog timer (or computer operating properly (COP) timer) is a computer hardware or software timer that triggers a system reset or other corrective action if the main program, due to some fault condition, such as a hang, neglects to regularly service the watchdog (writing a "service pulse" to it, also referred to as "kicking the dog", "petting the dog", "feeding the watchdog" or "waking the watchdog"). The intention is to bring the system back from the unresponsive state into normal operation. Watchdog timers can be more complex, attempting to save debug information onto a persistent medium; i.e. information useful for debugging the problem that caused the fault. In this case a second, simpler, watchdog timer ensures that if the first watchdog timer does not report completion of its information saving task within a certain amount of time, the system will reset with or without the information saved. The most common use of watchdog timers is in embedded systems, where this specialized timer is often a built-in unit of a microcontroller.

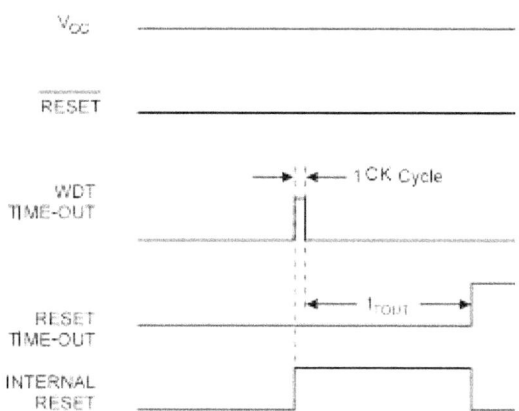

43. What is the difference between 8 bit 16 bit and 32 bit processor?

Different families of micros vary in their capabilities. The number of bits just refers to the width of the data pipe, which limits the precision of math, although many micros will either emulate higher order math or have special HW that can perform higher precision math functions.

The historic difference has been price: 8-bit was cheapest, 32-bit was expensive. This is still true in generally, but the price of 16-bit parts have come down significantly.

Most 8-bit processors are old and run on old architectures, so they tend to be slower. They are also made more cheaply, since that is where the competition is at the 8-bit point, and this makes them tend towards slowness. They also tend to have a low limit on supported RAM/other storage, but the actual amount depends on the family.

16-bit processors tend to focus on price as well, but there is a large range of parts available, some of which have fairly high performance and large amounts of on-chip peripherals. These parts usually perform faster than 8-bit parts on math where the precision is greater than 8 bits, and tend to have more addressable memory.

32-bit chips compete primarily on performance for an application. There is a considerable range of 32-bit parts available, each targeted at some specific application. They tend to come loaded with peripherals and compete on feature completeness. They have a large amount of addressable memory and the performance tends to be better than 16-bit parts.

44. What is a Function Pointer ?

A function pointer is a variable that stores the address of a function that can later be called through that function pointer. This is useful because functions encapsulate behavior. For instance, every time you need a particular behavior such as drawing a line, instead of writing out a bunch of code, all you need to do is call the function. But sometimes you would like to choose different behaviors at different times in essentially the same piece of code.

Example: int (*fp) (int, int); -> Function pointer returning an integer

45. Size of Datatypes

Name	Description	Size*	Range*
char	Character or small integer.	1byte	signed: -128 to 127 unsigned: 0 to 255
short int (short)	Short Integer.	2bytes	signed: -32768 to 32767 unsigned: 0 to 65535
int	Integer.	4bytes	signed: -2147483648 to 2147483647 unsigned: 0 to 4294967295
long int (long)	Long integer.	4bytes	signed: -2147483648 to 2147483647 unsigned: 0 to 4294967295
bool	Boolean value. It can take one of two values: true or false.	1byte	true or false
float	Floating point number.	4bytes	+/- 3.4e +/- 38 (~7 digits)
double	Double precision floating point number.	8bytes	+/- 1.7e +/- 308 (~15 digits)
long double	Long double precision floating point number.	8bytes	+/- 1.7e +/- 308 (~15 digits)

46. What is the difference between typedef & Macros?

Typedef is used to create a new name to an already existing data type. Redefine the name creates conflict with the previous declaration.

eg:
typedef unsigned int UINT32

Macros [#define] is a direct substitution of the text before compling the whole code. In the given example, its just a textual substitution. where there is a posibility of redefining the macro

eg:
#define chPointer char *
#undef chPointer
#define chPointer int *

Typedef are used for declarations when compare with macro

typedefs can correctly encode pointer types.where as #DEFINES are just replacements done by the preprocessor.

For example,

1. typedef char *String_t;
2. #define String_d char *
3. String_t s1, s2; String_d s3, s4;

s1, s2, and s3 are all declared as char *, but s4 is declared as a char, which is probably not the intention.

47. What is the difference between a macro and a function?

Macros are essentially shorthand representations of arbitrary sections of the source code, which makes the source code, while its (the macro template's) expansion replaces each of its presence prior to compilation. Whatever is there to do with Macros, it is done by the preprocessor, so that the source code is ready for compilation. Function is a calling routine, whence a large program is divided into separate portions, each portion doing a separate job, and proper calling of these portions in different places combines the works done by them into the required complete output. Thus functions have nothing to do with the preprocessing period. they are just compiled. To some extent function and macro is similar, for a macro can occasionally be invoked to perform a task that is generally entrusted to a function. But the similarity ends there.

The differences are:

1. Macro consumes less time. When a function is called, arguments have to be passed to it, those arguments are accepted by corresponding dummy variables in the function, they are processed, and finally the function returns a value that is assigned to a variable (except for a void function). If a function is invoked a number of times, the times add up, and compilation is delayed. On the other hand, the macro expansion had already taken place and replaced each occurrence of the macro in the source code before the source code starts compiling, so it requires no additional time to execute.
2. Function consumes less memory. While a function replete with macros may look succinct on surface, prior to compilation, all the macro-presences are replaced by their corresponding macro expansions, which consumes considerable memory. On the other hand, even if a function is invoked 100 times, it still occupies the same space. Hence function is more amenable to less memory requirements

48. What is inline function?

Inline function is the optimization technique used by the compilers. One can simply prepend inline keyword to function prototype to make a function inline. Inline function instruct compiler to insert complete body of the function wherever that function got used in code.

Advantages :-

1) It does not require function calling overhead.
2) It also save overhead of variables push/pop on the stack, while function calling.
3) It also save overhead of return call from a function.

4) It increases locality of reference by utilizing instruction cache.

5) After in-lining compiler can also apply intraprocedural optmization if specified. This is the most important one, in this way compiler can now focus on dead code elimination, can give more stress on branch prediction, induction variable elimination etc..

Disadvantages :-

1) May increase function size so that it may not fit on the cache, causing lots of cahce miss.

2) After in-lining function if variables number which are going to use register increases than they may create overhead on register variable resource utilization.

3) It may cause compilation overhead as if some body changes code inside inline function than all calling location will also be compiled.

4) If used in header file, it will make your header file size large and may also make it unreadable.

5) If somebody used too many inline function resultant in a larger code size than it may cause thrashing in memory. More and more number of page fault bringing down your program performance.

6) Its not useful for embeded system where large binary size is not preferred at all due to memory size constraints

49. What is the difference between a macro and a inline function?

Macros :

1. input argument datatype checking can't be done.

2. compiler has no idea about macros

3. Code is not readable

4. macros are always expanded or replaced during preprocessing, hence code size is more.

5. macro can't return.

Inline function :

1. input argument datatype can be done.

2. compiler knows about inline functions.

3. code is readable

4. inline functions are may not be expanded always

5. can return.

50. Preprocessor Statements #ifdef, #else, #endif

These provide a rapid way to "clip" out and insert code.

Consider:

```
#define FIRST




main()


{


int a, b, c;
```

```
#ifdef FIRST

    a=2; b=6; c=4;

#else

    printf("Enter a:");

    scanf("%d", &a);


    printf("Enter a:");

    scanf("%d", &a);


    printf("Enter a:");

    scanf("%d", &a);

#endif

    additonal code
```

Note that if FIRST is defined (which it is in the above) the values of a, b and c are hardcoded to values of 2, 6 and 4. This can save a lot of time when developing software as it avoids tediously typing everything in each and everytime you run your routine. When FIRST is defined, all that is passed to the compiler is the code between the #ifdef and the #else. The code between the #else and the #endif is not seen by the compiler. It is as if it were all a comment.

Once you have your routine working, and desire to insert the printf and scanfs, all that is required is to go back and delete the the #define FIRST. Now, the compiler does not see the;

```
    a=2; b=6; c=4;
```

How to calculate CRC Sequence in a CAN Frame?

The receivers calculate the CRC in the same way as the transmitter as follows:

1. The message is regarded as polynom and is divided by the generator polynom: $x^{15} + x14 + x10 + x8 + x7 + x4 + x3 + 1$.
2. The division rest of this modulo2 division is the CRC sequence which is transmitted together with the message.
3. The receiver divides the message inclusive the CRC sequence by the generator polynom.

A CRC error has to be detected, if the calculated result is not the same as that received in the CRC sequence. In this case the receiver discards the message and transmits an Error Frame to request retransmission.

51. Difference between static and dynamic RAM?
Static RAM (SRAM) – High cost & Fast

1. 4 times more expensive
2. Very low access time
3. Can store ¼ as much
4. Information stored on RS flip-flops
5. No need for refreshing

Dynamic RAM (DRAM) – Low cost & slow

1. Low cost
2. Consumes less power
3. Can store 4 times as much
4. Information stored on FET transistors
5. Needs to be refreshed

CANoe Tool Questions:
What is difference the between IG and G block in CANalyzer/CANoe tool?
Answer: There are two limitations to the Generator block that limit its effectiveness in complex tasks. The block is misleading for some people because it requires multiple windows for setting up the transmit message list. The second problem is the block settings have to be

set before the CANalyzer measurement starts. No changes can be made if the measurement is running.

Fortunately, CANalyzer has another transmission block that eliminates both practical limitations: the Interactive Generator block (**IG**). The **IG** block combines the configuration windows of the Generator block into one window; therefore, everything can be setup in one spot. In addition, changes can be made with the **IG**.

Without CAPL,can we simulate the other ECU's CAN Messages except Test ECU in the CAN Simulation Network in CANoe tool without using IG or G blocks.

How to change the baud rate in CANoe without changing the code?

The bit rate may be changed by either changing the oscillator frequency, which is usually restricted by the processor requirements, or by specifying the length of the bit segments in "time quantum" and the prescaler value.

In Canoe tool, we can change the bus timing register 0 & 1 values for correcting the baud rate.

In Autosar, we can use post build configuration for CAN baudrate values.

What is environment variable?

Environment variables are data objects global to the CANoe environment and are used to link the functions of a CANoe panel to CAPL programs.

What is HIL and SIL testing?

Answer: Hardware-in-the-loop (HIL) simulation is a technique that is used in the development and test of complex real-time embedded systems. HIL simulation provides an effective platform by adding the complexity of the plant under control to the test platform. The complexity of the plant under control is included in test and development by adding a mathematical representation of all related dynamic systems. These mathematical representations are referred to as the "plant simulation". The embedded system to be tested interacts with this plant simulation.

Hardware-In-the-Loop System is an effective platform for developing and testing complex real-time embedded systems. HIL system provides the complexity of the plant under control using mathematical representation, called "plant simulation", of all related dynamic systems. It also includes electrical emulation of sensors and actuators which act as the interface between the plant simulation and the embedded system under test.

Advantages of HIL System

- Provides Cost Savings by Shortened Development time
- Complete, consistent test coverage.
- Supports automated testing
- Enables testing the hardware without building a "plant prototype"
- Simulator performs test outside the normal range of operation
- Supports reproducible test runs that can assist in uncovering and tracking down hard to find problems.
- Enables testing with less risk of destroying the system

SIL: SIL refers to the kind of testing done to validate the behavior of the C-code used in the controller. That code can be auto-generated from the model used in algorithm development. Emmeskay has a deep understanding of SIL testing and auto-code generation from the many SIL projects we have performed for our customers.

Testing and Validation

- Plant model developed in vehicle simulation environment is imported to Simulink as a library.
- Controller is tested in loop with the plant for different routes and speed profiles.
- Controller is tested for different fault modes of the system using GUI VisualConnex

RTOS Question:

What is RTOS?

Real-Time Operating System is a multitasking operating system intended for real-time applications. It is used on every device/system needing real time operations that means operations based not only on correctness but also upon the time (clock cycles) in which they are performed.

In general, an operating system (OS) is responsible for managing the hardware resources of a computer and hosting applications that run on the computer. An RTOS performs these tasks, but is also specially designed to run applications with very precise timing and a high degree of reliability. This can be especially important in measurement and automation systems where downtime is costly or a program delay could cause a safety hazard. To be considered "real-time", an operating system must have a known maximum time for each of the critical operations that it performs (or at least be able to guarantee that maximum most of the time). Some of these operations include OS calls and interrupt handling. Operating systems that can absolutely guarantee a maximum time for these operations are commonly referred to as "hard real-time", while operating systems that can only guarantee a maximum most of the time are referred to as "soft real-time".

Example: Imagine that you are designing an airbag system for a new model of car. In this case, a small error in timing (causing the airbag to deploy too early or too late) could be catastrophic and cause injury. Therefore, a hard real-time system is needed; you need assurance as the system designer that no single operation will exceed certain timing constraints. On the other hand, if you were to design a mobile phone that received streaming video, it may be ok to lose a small amount of data occasionally even though on average it is important to keep up with the video stream. For this application, a soft real-time operating system may suffice. An RTOS can guarantee that a program will run with very consistent timing. Real-time operating systems do this by providing programmers with a high degree of control over how tasks are prioritized, and typically also allow checking to make sure that important deadlines are met.

How Real-Time OSs Differ from General-Purpose OSs?

Operating systems such as Microsoft Windows and Mac OS can provide an excellent platform for developing and running your non-critical measurement and control applications. However, these operating systems are designed for different use cases than real-time operating systems, and are not the ideal platform for running applications that require precise timing or extended up-time. This section will identify some of the major under-the-hood differences between both types of operating systems, and explain what you can expect when programming a real-time application.

Interrupt Latency

Interrupt latency is measured as the amount of time between when a device generates an interrupt and when that device is serviced. While general-purpose operating systems may take a variable amount of time to respond to a given interrupt, real-time operating systems must guarantee that all interrupts will be serviced within a certain maximum amount of time. In other words, the interrupt latency of real-time operating systems must be bounded

Unanswered Interview Questions :(If you know comment as reply)
What is the use of Passive error node?
Error Passive receivers can no longer interrupt the data transfer as a recessive Error Flag does not influence the bus levels. An Error Passive transmitter can still interrupt its own message by sending a passive Error Flag. Attention, if one Receiver is in error passive mode no data consistency is guaranteed any more.

How to find the bug in code using debugger if pointer is pointing to a illegal value?
If two CAN messages with same ID sending at a same time, different data which can node will gain arbitration? How to test it?
Is it possible to declare struct and union one inside other? Explain with example
  1. Spi and I2C difference.?
  2. What is UDS advantages?
  3. What is cross compiler
  4. Unit/integration/all testings.
  5. Regression testing.
  6. Test case types.
  7. Malloc calloc
  8. Function pointers Advantage where it is used?

Hope these questions help you in Interview.